大学生人工智能素养红皮书

（2024年版）

Red Book on Artificial Intelligence Literacy of College Students

(2024 Edition)

浙江大学人工智能教育教学研究中心　编著

Zhejiang University Research Center of
Artifical Intelligence for Education and Teaching

·杭州·

图书在版编目（CIP）数据

大学生人工智能素养红皮书：2024年版 / 浙江大学人工智能教育教学研究中心编著. -- 杭州：浙江大学出版社，2024.8.（2025.3 重印）-- ISBN 978-7-308-25593-6

Ⅰ.TP18

中国国家版本馆CIP数据核字第2024YS2736号

大学生人工智能素养红皮书：2024年版
DAXUESHENG RENGONG ZHINENG SUYANG HONGPISHU: 2024 NIAN BAN

浙江大学人工智能教育教学研究中心　编著

策划编辑	黄娟琴　李　晨
责任编辑	柯华杰　陈丽勋
责任校对	郑成业
封面设计	林智广告
出版发行	浙江大学出版社
	（杭州市天目山路148号　邮政编码310007）
	（网址：http://www.zjupress.com）
排　　版	杭州林智广告有限公司
印　　刷	杭州捷派印务有限公司
开　　本	710mm×1000mm　1/16
印　　张	6.5
字　　数	85千
版 印 次	2024年8月第1版　2025年3月第2次印刷
书　　号	ISBN 978-7-308-25593-6
定　　价	36.00元

版权所有　侵权必究　　印装差错　负责调换

浙江大学出版社市场运营中心联系方式：0571-88925591；http://zjdxcbs.tmall.com

前　言

　　与任何有机体一样，人类需要与环境保持良好的互动，才能更好地繁衍生息。在一般意义上，素养指的是人类个体与环境互动的能力，它旨在维持和促进个体与种群的生存和发展，具有主体性、实践性和伦理性等内在特征。

　　先前的技术发明从机械化增强角度提升了人类与环境的互动能力，然而，人工智能的出现挑战了人类的根本，它深刻改变了人类与环境互动的能力和角色，成为引领新一轮科技革命和产业变革的战略性技术，对经济发展、社会进步、全球政治经济格局以及教育变革产生着重大而深远的影响。

　　在教育领域，人工智能正彻底改变着以知识积累和传递为中心的教学模式，新一代人工智能技术可以胜任原本属于人类脑力劳动范畴的部分工作。如何通过多样化的人机协同模式不断提升学习者的主体性、能动性和实践性，最终让大规模的个性化学习普遍发生，成为"智能时代，教育何为"需要直面的重大命题。

生成式人工智能将人类绝大多数知识装进数字化知识容器中，重构了人类知识版图，成为与人类并驾齐驱的"知识提供者"。然而，生成式人工智能的局限性也是显而易见的，它对个体自主思考、判断、学习能力乃至伦理道德观提出了前所未有的挑战。在人类教育史上，每一次伟大的技术创新（如文字、印刷术和互联网技术）都引发了教育领域质和量的飞跃。以 ChatGPT 为代表的生成式人工智能技术将实现对传统教育体系的再次迭代升级，促使人类教育目标从知识本位和能力本位走向素养本位。

本红皮书旨在提出大学生人工智能素养的构成内涵、培养的目标与愿景以及培养的载体、行动与策略，认为大学生人工智能素养是由体系化知识、构建式能力、创造性价值和人本型伦理构成的有机整体，其中知识为基、能力为重、价值为先、伦理为本。

PREFACE

Like any other organism, humans need to interact well with their environment to live and thrive. In a general sense, literacy refers to the ability of individuals to interact with the environment. It aims to maintain and promote the survival and development of individuals and population, and is inherently characterized by subjectivity, practicality, and ethics.

Previous technological inventions have improved the ability of human beings to interact with the environment from the perspective of mechanized enhancement. However, the emergence of artificial intelligence (AI) has challenged the fundamentals of human beings. AI has profoundly transformed the ability and role of humans in interacting with the environment and has become a strategic technology leading to a new round of sci-tech revolution and industrial transformation. It has a significant and far-reaching impact on economic development, social progress, global political and economic patterns, and educational reform.

In the field of education, AI is completely changing the teaching model centered on knowledge accumulation and transfer. The new generation of AI is competent for some tasks that originally demanded human mental labor. Enhancing learners' subjectivity, initiative and practicality through diversified human–computer collaboration models to enable large-scale personalized learning, is a major challenge that we face today. We need to discuss: "What is the purpose of education in the intelligent era?"

Generative artificial intelligence (GenAI) puts most of human knowledge into digital knowledge containers, reconstructs the human knowledge landscape, and becomes a "knowledge provider" that keeps pace with humans. However, GenAI shows obvious limitations. It poses unprecedented challenges to individual independent thinking, judgment, learning ability and even ethics. Throughout the history of human education, every great technological innovation (such as writing, printing and the internet) has triggered a qualitative and quantitative leap in the field of education. GenAI, represented by ChatGPT, will achieve another iterative upgrade of the traditional education system, promoting the goals of human education from one based on knowledge and skills to one based on literacy.

This Red Book aims to propose the components of AI literacy among college students, along with the objectives and aspirations for cultivation. It also delves into the methods, actions, and strategies necessary for fostering such literacy. This Red Book proposes that the AI literacy of college students is an organic whole composed of four factors: systematic knowledge, constructive capability, creative value, and humanistic ethics. Among these four factors, systematic knowledge lays the bedrock, constructive capability forms the core, creative value lights the path ahead, and humanistic ethics remain the essence evermore.

目　录

第一部分
人工智能发展给高等教育带来的挑战

一、从达特茅斯启航到新一代人工智能 ······ 003

二、人工智能带给高等教育的挑战和机遇 ······ 004

第二部分
大学生人工智能素养培养的目标与愿景

一、人工智能素养的概念及构成内涵 ······ 011

二、大学生人工智能素养培养的目标 ······ 013

三、大学生人工智能素养培养的愿景 ······ 015

第三部分
大学生人工智能素养培养的载体、行动与策略

一、大学生人工智能素养培养的载体 ······ 021

二、大学生人工智能素养培养的行动 ······ 024

三、大学生人工智能素养培养的策略 ······ 025

结　语 ······ 028

附 录

附录1：教育部计算机"101计划"核心课程"人工智能引论"知识点体系 029

附录2：ACM和IEEE-CS制定的新版人工智能知识点 ... 032

附录3：联合国教科文组织发布的《K-12 AI课程：政府认可的AI课程图谱》 036

CONTENTS

Part I

Challenges Posed to Higher Education by the
Advancement of Artificial Intelligence

    Sailing from Dartmouth to the New Generation Artificial Intelligence 043

    Challenges and Opportunities of Artificial Intelligence for Higher Education 045

Part II

Objectives and Vision of Artificial Intelligence Literacy
Cultivation for College Students

    Concept and Constitutive Connotation of Artificial Intelligence Literacy 053

    The Goal of Cultivating College Students' Artificial Intelligence Literacy 055

    Visions of Artificial Intelligence Literacy Cultivation for College Students 059

Part III
Carriers, Actions and Strategies for Artificial Intelligence Literacy Cultivation for College Students

    Carriers for Artificial Intelligence Literacy Cultivation for College Students 066

    Actions for Artificial Intelligence Literacy Cultivation for College Students 070

    Strategies for Artificial Intelligence Literacy Cultivation for College Students 073

Conclusion 077

Appendices

    Appendix 1   Syllabus of Introduction to Artificial Intelligence, one of the core courses of the 101 Plan for Computer Education implemented by the Ministry of Education (China) 079

    Appendix 2   The new version of artificial intelligence knowledge areas formulated by the ACM and IEEE-CS 084

    Appendix 3   UNESCO K-12 AI curricula: A mapping of government-endorsed AI curricula 089

第一部分

人工智能发展给高等教育带来的挑战

一、从达特茅斯启航到新一代人工智能

1955年8月，时任达特茅斯学院数学系助理教授、1971年度图灵奖获得者约翰·麦卡锡（John McCarthy），时任哈佛大学数学系和神经学系青年研究员、1969年度图灵奖获得者马文·李·明斯基（Marvin Lee Minsky），时任贝尔实验室数学家、信息论之父克劳德·香农（Claude Shannon）和时任IBM信息研究主管、IBM第一代通用计算机701主设计师纳撒尼尔·罗切斯特（Nathaniel Rochester）四位学者向美国洛克菲勒基金会递交了一份题为《关于举办达特茅斯人工智能夏季研讨会的提议》的建议书[①]，希望洛克菲勒基金会资助拟于1956年夏天在达特茅斯学院举办的人工智能研讨会。在这份建议书中，"人工智能（artificial intelligence, AI）"这一术语首次被使用。1956年6月18日至8月17日，近40位学者来到美国达特茅斯学院参加人工智能研讨会，围绕自动计算模拟人脑高级功能、使用通用语言进行计算机编程以模仿人脑推理、神经元相互连接形成概念、对计算复杂性的度量、算法

---

① *John McCarthy, Marvin Minsky, Nathaniel Rochester, Claude Shannon: A proposal for the Dartmouth summer research project on artificial intelligence, 1955*

自我提升、算法的抽象能力、随机性与创造力等议题展开讨论，开启了人类研发人工智能之路。

在之后70年的发展历程中，以神经网络为代表的连接主义，以知识工程为代表的符号主义和以控制论为代表的行为主义相互渗透发展。21世纪初，以"深度学习"为代表的人工智能方法在计算机视觉、语音识别和游戏博弈等领域取得显著进展。

中国人工智能发展与时俱进。2017年，国务院发布了《新一代人工智能发展规划》，其中提出了大数据智能、跨媒体智能、群体智能、人机混合增强智能、智能自主系统等五大智能形态，指出人工智能呈现深度学习、跨界融合、人机协同、群智开放、自主操控等新特征，标志着中国人工智能发展进入新阶段。

2022年底出现的大语言模型计算范式推动了人工智能从"一个模型解决一个任务"迈向"一个模型解决所有任务（all in one）"的新计算架构发展阶段。这一架构的核心就是生成式人工智能，它以强大的内容合成能力为特征，推动了语言生成和对话式人工智能等领域的突破性进展。生成式人工智能的发展将进一步推动人工智能技术的普及和深入应用，为社会带来更多的便利和创新。

二、人工智能带给高等教育的挑战和机遇

和历史上蒸汽机、电力、计算机和互联网等发明一样，人工智能正在成为一种通用目的技术，以史无前例的速度全方位地改

变着人类社会的发展。根据世界经济论坛发布的《未来就业报告2023》,到2025年,人工智能将导致全球减少8500万个工作岗位,同时也将创造9700万个新的工作岗位,净增加1200万个工作岗位[①]。该报告同时指出,人工智能与大数据能力是未来最需要培养的十项技能之一。鉴于人工智能行业应用的普遍性及对未来社会发展的重要性,联合国教科文组织(UNESCO)在2023年提出的"在校师生人工智能能力框架"将人工智能素养列为学生及教师必备素养[②],指出在校师生需要掌握人工智能相关的知识、技能和态度,在教育及其他领域可以通过安全且有意义的方式理解和使用人工智能。由人工智能引发的行业用人需求以及职场技能要求变化正在倒逼高等教育人才培养方向和方式的优化与革新。

(一)人工智能引发知识生产模式变革

通过智能算法,特别是Transformer[③]这样的神经网络架构,人工智能能够从海量语料中学习单词与单词之间的共生关联关系,实现自然语言的合成。以Transformer为核心构建的ChatGPT等生成式人工智能系统通过洞悉海量数据中单词—单词、句子—句子等之间的关联性,按照规模法则(scaling law)不断增大模型规模和超越"费曼极限"增强模型非线性映射能力,迅速具备了

---

① https://www.weforum.org/publications/the-future-of-jobs-report-2023/
② https://www.unesco.org/en/digital-education/ai-future-learning/competency-frameworks
③ Vaswani, Ashish, Noam Shazeer, Niki Parmar, Jakob Uszkoreit, Llion Jones, Aidan N. Gomez, Łukasz Kaiser, and Illia Polosukhin: Attention is all you need, Proceedings of Advances in Neural Information Processing Systems, 2017

合成语言的能力，犹如昨日重现一样对单词进行有意义的关联组合，连缀成与场景相关的会意句子，生成有价值的句子和知识。

通过其强大的自然语言处理和生成能力，生成式人工智能不仅可以根据用户的需求和偏好生成个性化的内容，还能够处理和理解海量的语料库，这意味着知识的生产不再完全依赖于人类的个体能力和时间成本，而是可以通过算法实现高效且大规模的生产。这种变革不仅提高了知识生产的效率和速度，还为人类知识的整合、传播和创新提供了全新的可能，推动着高等教育朝着更加智能化和信息化的方向发展。

（二）人工智能引发高校课堂教学模式变革

全球很多高校在人工智能课堂应用领域已经开展了有益的探索。如果利用得当，人工智能将为学生和教师提供更多个性化、交互式和创新性的学习方式和学习内容：虚拟助教、聊天机器人、智能测评等工具正在助力教学提效增质；生成式人工智能教育应用将推动教学模式从"师—生"二元结构转向"师—机—生"三元结构，推动学习空间泛在化，满足学习过程全覆盖的个性化需求，创建人机协同的学习空间。引入"人在回路"的闭环协同学习机制，形成数据驱动下归纳、知识指导中演绎以及反馈认知中顿悟等相互结合的计算理论模型，将是未来"师—机—生"耦合而成的学习空间中"教"与"学"的发展方向。时时、处处、人人可学的泛在学习将更加普及，因材施教的千年梦想将成为可能。

（三）人工智能引发高校科研范式转型

人工智能正全面融入科学、技术和工程研究，帮助研究者生成假设、设计实验、计算结果、解释机理，特别是辅助研究者在不同的假设条件下进行大量重复的验证和试错，人机协同模式大大加速了科学创新的进程，探索先前无法触及的知识视野和领域天地。大语言模型以自然语言形式与人类交互，同时将各种应用以插件形式进行整合，成为链接信息空间—物理世界—人类社会三元空间的流量入口。此外，智能体作为能够感知自身环境、自我决策并采取行动的人工智能模型，与生成式人工智能基座模型相结合，形成了人工智能体（AI agents）这一垂直领域前沿。人工智能体在内容合成的基础上，能够实现信息检索、人机对话、任务执行、逻辑推理等自主行为。将个人智能体应用于科研领域可以辅助学习者探索生成假设、设计实验、计算结果、解释机理等步骤，促进了人类学习能力的提升，也拓展了未知空间的研究范式。

（四）人工智能在高校应用中可能产生的负面影响

生成式人工智能的产生使得智能机器成为知识生产的辅助者，对个体学习者的自主思考、判断、学习能力乃至伦理道德观都提出了挑战。如果使用不当，人工智能的教育应用也会带来很多负面影响，如教师地位边缘化、学生学习"孤岛"化、知识体系碎片化、隐私泄露风险、歧视和偏见、伦理风险、学术诚信和公平失衡、教育关系异化、安全和隐私问题、知识盲区与信息茧房、内容准确性不足、学生高阶思维被削弱、数字应用鸿沟等。新一代人工智能技术具有深度学习、跨界融合、人机协同、群智开放、自主操控等特征,其可解释性低、系统偏差、数据安全、数据隐私等问题也给行业和社会带来了前所未有的伦理风险与挑战。

"智能时代,高教何为?"中国高校教师应准确识变、科学应变、主动求变,重构智能时代中国高校各专业的人才培养目标、路径以及支持系统,为社会培养大批量具有人工智能素养的复合型专业人才,助力未来社会朝着和谐、健康、可持续的方向发展。

2

第二部分

大学生人工智能素养培养的目标与愿景

一、人工智能素养的概念及构成内涵

中文"素养"一词出自《后汉书·卷七四下·刘表传》,"越有所素养者,使人示之以利,必持众来"。它指的是平日的修养,广义上它包含道德品质、外表形象、知识水平与能力等各个方面。

1997—2005年,经济合作与发展组织(Organization for Economic Cooperation and Development,OECD)依托"素养的界定与遴选:理论和概念基础"项目,围绕"素养"开展了为期九年的研究,最终将素养界定为"在特定情境中,通过利用和调动心理社会资源(包括技能和态度),以满足复杂需要的能力"[1],同时指出,素养具有时代性、整体性、发展性及可测性。即素养依存于特定情境,凡是有助于个体适应社会或解决复杂问题的能力与技巧,都可称为素养;素养是知识、能力与态度的统一;通过特定教育手段,能够实现素养的培养;素养能够通过可理解、可操作、可评估的指标进行度量。

---

[1] Marilyn Mathieu: The definition and selection of key competencies: Executive summary, 2005

2018年，UNESCO将素养定义为"使用与情境相关的所有材料进行识别、理解、解释、创造、交流和计算的能力"。

可见，个人在特定情境中，面对生活中复杂多变的问题时，能利用和调动知识、技能、判断等能力，观察和理解世界、构建解决问题的方案、用行动去检验对世界的认识是否合理，提升自身胜任力，创造人类增量知识，这种综合表现可被称为"素养"。

国际计算机学会（ACM）和电气与电子工程师协会计算机分会（IEEE-CS）在2020年版《计算课程体系规范》（Computing Curricula）中用"胜任力"这一概念来描述结合知识、技能和品行三个维度内容的计算机专业素养。

随着人类进入智能时代，人工智能素养（AI literacy）逐渐成为个体生存和发展的重要素养之一。这一概念的首次提出是在20世纪70年代[①]，当时主要强调的是人工智能专业技术人员的素养组成。随着人工智能对人类社会产生的巨大影响，每位公民都需要了解人工智能是什么、人工智能可以做什么和不可以做什么、如何负责任地使用人工智能以及质疑人工智能的使用，让人工智能为个体、群体和人类公共利益服务。

近年来，UNESCO一直重视全球的人工智能教育，认为所有公民都需要具备包含知识、理解、技能和价值观等要素的"人工智能素养"——这已成为21世纪的基本语法。

---

① Agre, Philip E.: *What to read: A biased guide to AI literacy for the beginner*, MIT Artificial Intelligence Laboratory Working Papers, WP-239, 1972

二、大学生人工智能素养培养的目标

作为高等教育的主要培养对象以及社会各行各业未来的接班人，大学生是一个备受各国社会关注和期待的群体。整体而言，大学生年富力强，精力充沛，学习能力和学习习惯基本形成，具有极强的可塑性、创造力及发展潜力。正因为大学生群体的一些特质及国家和社会对该群体的特别期待，智能时代有关大学生人工素养的内涵也有别于其他社会群体。

为了应对人工智能给社会各行各业所带来的前所未有的挑战，高校需要培养大学生在人工智能时代具备了解人工智能、使用人工智能、创新人工智能和恪守人与人造物关系等综合能力，提升他们的人工智能素养，让他们能够在智能社会中更美好地生活与发展。

具体而言，大学生人工智能素养包括体系化知识、构建式能力、创造性价值和人本型伦理四个有机整体，其中，知识为基、能力为重、价值为先、伦理为本，四者相辅相成、相互融合。

体系化知识：认知是人类智能的重要表现，其基石是体系化知识，体系化知识意味着可对学习对象进行整体性理解和系统化分类。人工智能具有至小有内、至大无外的交叉渗透特点，掌握了体系化知识就可以更清晰地认识到人工智能的内涵、边界和外延。

体系化知识是一个相对和动态的概念。人类思维的根本任务之一就是对纷繁复杂的事物进行分类，逐步形成系统化和综合性思维，渐次提升通用认知，达成共识。

构建式能力： 人工智能可在人和机器之间建立合作关系，统筹人工智能和人类智能的各自优势，共同努力实现特定任务的目标。在人机协同过程中，人类从数据中获得更多洞见，并确定最优解决方案，以前所未见的辅助方式完成任务。

长久以来，科学遵循着从假设到实验再到理论验证的循环，其核心在于寻求现象背后的可解释原理。在人工智能时代，人们可通过使用人工智能工具，构建人在回路闭环中解决问题的能力：对问题进行抽象建模、生成可验证假设、设计可计算模型、解释算法运行结果，根据反馈不断通过枚举和仿真等方式优化求解方法。构建式能力克服了传统方法难以驾驭数据复杂性的不足，推动从"知识本位教育"向"能力本位教育"转变。

创造性价值： 生成式人工智能对人类所有语料上下文信息进行压缩，然后概率合成，其对已有知识记忆和整合的强大能力使得以知识积累为中心的教育模式优势荡然无存。

通过人工智能增强主体性、彰显个性化、放大能动性和参与增强实验，产出人类增量知识，形成创造性价值，进而成为社会所共同积累和分享的"普遍智能"。

人本型伦理： 传统的科技发展往往采取一种所谓的"技术先行或占先行动路径（proactionary approach）"模式，以发展技术为优先原则，体现出一种强大的工具理性，即"通过缜密的逻辑思维和精细的科学计算来实现效率或效用的最大化"。

随着物联网、人工智能等技术的发展，人类已经完全进入一个由"信息空间—物理世界—人类社会"构成的三元空间结构之中,其中的伦理学讨论不再只是人际关系,也不是人与自然界既定事实之间的关系,而是人类与人造物在社会中所构成的关联,使得人工智能具有技术和社会双重属性。

因此,人机共融社会中,人类应遵守以人为本、智能向善的伦理理念,确保把人类价值观、道德观和法律法规贯穿于人工智能的产品和服务,赋予人工智能社会属性。图1描述了大学生人工智能素养的内涵、培养载体、行动及策略。

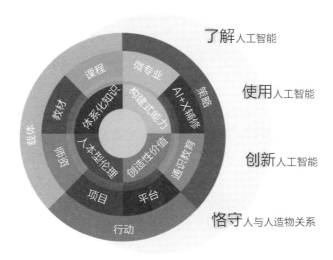

图1 大学生人工智能素养的内涵、培养载体、行动及策略

三、大学生人工智能素养培养的愿景

大学生人工智能素养培养的四大愿景包括形成人工智能思辨模式、具备人工智能解决问题的能力、创造人类增量知识的价值、坚持以人为本的伦理底线。表1从知识、能力、价值和伦

理四个构成维度给出了大学生人工智能素养的具体内容及培养愿景。

表1 大学生人工智能素养的构成、内容及愿景

构成	内容	愿景
体系化知识	●数据与知识：人工智能之燃料 ●算法与模型：人工智能之引擎 ●算力与系统：人工智能之载体 ●交叉与应用：人工智能之用途 ●可信与安全：人工智能双刃剑	形成人工智能思辨模式：人工智能的"能"与"不能"相对转变、人工智能中的确定性（逻辑）和不确定性（概率）辩证统一、机器智能与自然智能的共生协同、"人工智能+"学科交叉与综合、科技属性与社会属性高度融合
构建式能力	●对求解问题的抽象和建模能力 ●对求解过程的分解和模块化能力 ●对求解方法的可验证假设能力 ●对求解结果的解释反馈能力 ●利用生成式人工智能求解问题能力	具备人工智能解决问题的能力：培养设计与构造的计算思维，机器智能归纳和人类智能直觉等融通共进，塑造通过人机协同机制进行解决问题的构建能力，实现从"知识本位教育"向"能力本位教育"转变
创造性价值	●目标引导式对话下的内容重构 ●师—机—生交互中的认知主体性增强 ●个性化学习体验的自主性融入 ●解决问题的实践能动性体验 ●克服依赖智能工具的选择性自省	创造人类增量知识的价值：在人工智能辅助下提升个性化、主体性和能动性，通过内容重构合成、实践探索、交互认知等手段创造价值，实现从"知识学习、能力塑造"向"价值创造"转变

续表

构成	内容	愿景
人本型伦理	● 数据安全与隐私保护的意识 ● 算法偏差与模型幻觉的警惕 ● AI 向善和以人为本的对齐 ● 人机共生共融的 AI & All 理念 ● 人类累积知识普惠共享的追求	坚持以人为本的伦理底线：从数据、算法、模型和应用等方面知晓人工智能脆弱性所带来的潜在危害，理解 AI 向善和以人为本的对齐模式，树立人机和谐相处和普惠智能的 AI & All 理念

体系化知识包括人工智能之燃料、引擎、载体和用途有关的数据与知识、算法与模型、算力与系统、交叉与应用等知识，以及人工智能双刃剑所需要的可信与安全知识，掌握体系化知识才能具备应对复杂问题的综合系统思维。

图 2 展示了教育部计算机领域本科教育教学改革试点工作计划（"101 计划"）核心课程"人工智能引论"以"厚基础、强交叉、养品行、促应用"为理念所形成的体系化知识（详见附录 1）。

构建式能力包含在人机协作模式下有关求解问题的抽象和建模、分解和模块化、可验证假设、解释反馈等能力，以及利用生成式人工智能求解问题能力。

创造性价值包括目标引导式对话下的内容重构、师—机—生交互中的认知主体性增强、个性化学习体验的自主性融入、解决问题的实践能动性体验、克服依赖智能工具的选择性自省等创造

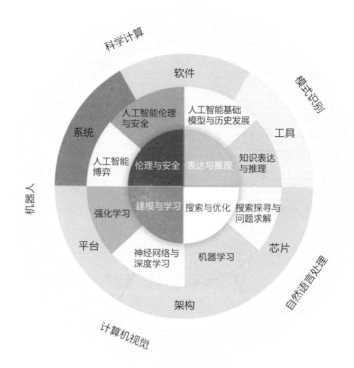

图2 教育部计算机"101计划"核心课程"人工智能引论"体系化知识

性价值模式。

人本型伦理包括数据安全与隐私保护的意识、算法偏差与模型幻觉的警惕、AI向善和以人为本的对齐、人机共生共融的AI & All理念、人类累积知识普惠共享的追求等。

第三部分

大学生人工智能素养培养的载体、行动与策略

一、大学生人工智能素养培养的载体

（一）课程：分类分级课程设置满足多样需求

在高校中，大学生人工智能素养培养最重要的载体之一就是系列课程的学习。针对不同专业及层次大学生的实际要求，高校可设置通识课程、专业课程、学科交叉课程、微课程等多类别课程，应对不同类型学生对人工智能的知识需求。表2给出了不同类别人工智能课程及涵盖的知识内容。

表2　高校多类别人工智能课程体系和内容描述

课程类别	课程内容描述
通识课程	介绍人工智能的历史、定义、分支、应用和前沿发展；讨论人工智能的伦理问题、隐私保护和安全挑战；探讨人工智能对未来社会的影响，包括就业、教育、医疗等
专业课程	讲授感知（如语音识别、自然语言理解、计算机视觉）、问题求解（如搜索和规划）、行动（如机器人）以及支持任务完成的体系架构（如智能体和多智能体）等不同方面的课程

续表

课程类别	课程内容描述
AI+X 交叉课程	介绍人工智能基本知识以及利用人工智能解决本学科问题的方法，如人工智能赋能经济、法律、艺术等不同学科；通过解决不同学科场景问题，明了人工智能对不同学科所产生的范式革命
微课程	以知识点为核心，围绕具体问题介绍人工智能不同知识点的内容，提供个性化学习的课程

（二）教材：数字化教学资源方便知识的传播

数字教材作为数字技术与教育教学深度融合的新形态教材，具有集成度高、互动性强、结构性明显等优点，为高校教育教学提供更加优质丰富的数字化资源，对加快教育数字化转型、推进教育强国建设具有重要意义。数字教材不仅是传统纸质教材的数字版，更是一种将教学内容、富媒体教学资源、学习工具和技术平台进行有机融合的新形态教材，为大规模"因材施教"提供可能。

数字教材在强化音视频的信息传输优势、极大地拓展读者阅读边界的同时，还可以有效聚焦读者目光、提升学习专注力、升级阅读体验，为学生个性化地"学"和教师创造性地"教"提供燃料。

人工智能教材所对应的视频、音频、实训案例和课件等教学资源可按照知识点组织，形成以"知识图谱—能力图谱"为结构的组织方式，为"学生更加主动地学、教师更加创造性地教"开辟广阔天地。

（三）师资：多元跨界融合促进教师团队建设

教师是加快教育数智化转型的最核心要素和关键群体，是促进人工智能技术与教育教学深度融合的执行者，需要通过多元跨界融合促进教师合作交流。

吸引跨学科教师加入，构建开放的人工智能师资团队： 人工智能是一个高度综合的交叉学科，涉及计算机科学、数学、神经科学、心理学等多个领域。为了更好地培养学生的人工智能素养，高校应积极吸引不同学科背景的教师参与人工智能课程建设。一方面，鼓励计算机、自动化、电子信息等相关专业的教师转型进入人工智能教学领域。学校可以为这些教师提供必要的培训和支持，帮助他们快速掌握人工智能的前沿知识和教学方法。同时，也要引导这些教师将人工智能与本专业深度融合，开发出交叉学科特色鲜明的课程。另一方面，要大力引进数学、物理、生物、医学、经济管理、社会学、艺术设计等其他学科的优秀教师，参与人工智能通识教育和交叉学科教学。

加强校企合作，引入企业专家参与教学： 人工智能是一个与产业联系紧密的应用导向型学科，加强与行业企业合作，引入企业专家和工程师参与教学，可以增强教学内容的前沿性和实践性。高校可以与人工智能领域的龙头企业建立长期合作关系，共建联合实验室、实践教学基地等。企业可以派遣优秀的专家和工程师到学校担任兼职教授，承担部分专业课程或实践环节的教学任务。

引导青年教师加入人工智能课程组：青年教师是高校教师队伍的生力军，在人工智能教育中可以发挥重要作用。学校应采取措施，引导和支持更多具有人工智能专业背景及对人工智能感兴趣的优秀青年教师加入人工智能课程教学团队。

二、大学生人工智能素养培养的行动

（一）汇项目：多元化设计促进理论与实践结合

大学应该建立多元化、开放性的人工智能实训项目体系，聚焦前沿技术和应用场景，促进学科交叉和校企协同，引导学生在实践中提升知识、能力、价值理念，尤其是强化伦理意识。通过参与伦理评估、跨界创新、开源贡献、产学研合作、社会服务等实践项目，学生可以深刻认识到人工智能的社会影响力，学会在技术创新和伦理规范间权衡取舍，在开放协作中贡献价值，在服务社会中担当责任。唯有将知识学习与实践应用紧密结合，才能真正培养出具备家国情怀、全球视野、创新能力和伦理素养的新时代人工智能领军人才。

大学可设立人工智能伦理、人工智能+X跨学科、人工智能开源生态、人工智能产学研、人工智能社会服务等实践项目，从不同方面加强大学生实训能力培养。

（二）建平台：搭建数字平台促进实践能力提升

全面数字化的教学资源建设： 教育数字化是开辟教育发展新赛道和塑造教育发展新优势的重要突破口，为个性化学习、终身学习、扩大优质教育资源覆盖面和教育现代化提供有效支撑。将体系化知识中每个知识点所对应的课程内容、教学微视频、实训题目等教学资源进行有机组织，构建数字化形态教学资源，创新知识载体，推动教育资源的供给侧结构性改革，以便让学生个性化地"学"、教师创造性地"教"。

生成式基座支持的学习平台： 生成式人工智能表现出较强的内容合成能力，推动了语言生成和对话式人工智能等领域的突破性进展，让人们看到技术变革对教育产生的重大影响。根据不同学科特点，通过打造基座、训练模型、调优模型、端侧智能体赋能和应用迭代等研制教育垂直领域大模型，赋能教育教学、科学研究和教育治理，发挥教育基础性、先导性、全局性作用，从支撑教育教学，到支撑科学研究，再到赋能千行百业的智能升级。

三、大学生人工智能素养培养的策略

（一）人工智能通识教育：素养养成的基石

教育教学将由教育者为引领转向以学习者为中心，专业发展转向通、专、跨的连贯发展。在智能时代，高校要将人工智能作为通识教育的重要内容。

人工智能课程要面向高校全体学生，普及人工智能基本概念，介绍人工智能发展现状，剖析人工智能未来趋势，引导大学生思考人工智能带来的社会变革，培养人文情怀和使命担当。

（二）AI+X辅修：促进交叉人才培养

为培养具备跨学科融合创新能力的高层次人工智能复合型人才，大学应积极探索AI+X纵向交叉人才培养模式。以人工智能或计算机专业人工智能方向本科生为基础，引导并支持其在研究生阶段跨专业学习，拓宽知识视野，提升创新实践能力。

具体而言，可采取**AI+X本博贯通、交叉学科课程、交叉科研训练、毕业设计交叉导师制、柔性学分互认机制、产学研用协同育人和国际交流合作等方法开展交叉人才培养**。通过在人工智能专业本科生教育中注重交叉学科素养的培育，为学生在研究生阶段进一步深化跨学科学习奠定良好基础。同时，要为学生提供多样化的跨学科学习资源和实践机会，激发其探索交叉领域的兴趣和潜力。高校应根据自身特色和优势，因地制宜地探索AI+X复合型创新人才培养的有效模式，为智能社会发展源源不断地输送高水平人才。

（三）微专业：跨专业协同促进AI+X人才培养

微专业由特定领域的一组微课程组成。微型课程可以帮助学习者在短时间内获得相对独立完整的单元知识。因此，微课程为专业人士提供了在不离开其现有岗位的情况下扩展其能力的潜在机会，促使他们的专业知识与时俱进。

微专业可打破院系和专业学科壁垒，联动政校企力量，汇聚一流的学者与产业专家共同开设课程，实现跨学院、跨学科、跨专业教学与管理，使非计算机专业和非人工智能专业的学生能够更为灵活、高效地学习和了解人工智能基本知识体系，掌握人工智能基本知识，提升人工智能实践应用能力，从而推动本学科今后研究的范式变革。

微专业可以采取多种灵活多样的方式，充分发挥跨学科交叉融合的特点，培养学生的人工智能素养和创新实践能力。微专业强调传统学科主动与AI融合，具体举措可以包括**校际联盟合作办学、跨院系协同开发课程、线上线下混合式教学、产教融合协同育人等**。

总之，高校开设人工智能微专业要体现交叉融合、开放协同的理念，打破学科专业壁垒，促进校校、校企、国际合作，共建共享优质教育资源。同时，要创新教育教学模式，线上线下结合，注重实践创新能力培养，多措并举提升人才培养质量。只有不断探索创新，人工智能微专业才能成为复合应用型人才培养的有效途径，更好地服务经济社会发展。

结　语

　　教育是把人从自然人转化为社会人的过程。在这一过程中，人类需要与环境保持持续互动，以维持生存和发展，这就突出了素养所涉及的知识、能力、价值和伦理问题。

　　在《说文解字》中，"教"和"育"分别指"上所施下所效"和"养子使作善"。人工智能时代对教育方法和手段产生了深刻影响。以ChatGPT为代表的大模型技术对教育进一步产生了巨大影响，使得机器和系统等人造物不再仅仅是知识的载体和表现工具，更是参与教与学的过程并成为其中的一方。

　　教育的首要目标永远应该是独立思考和判断的总体能力的培养，而不是获取特定的知识。为此，我们应该用科学理性的态度来把握技术的本质特征及潜力，对技术进行综合审视和哲学思考，对技术突破保持关注与警觉，更好地理解和发掘技术在教育领域的潜在价值。

　　人工智能能否实现伦理正当性，并不取决于人工智能本身，而是取决于人和人工智能是否真正指向正当性目的。面向未来，面对不确定性，我们的态度应该是从知识本位教育迈向能力本位教育，恪守人工智能发展伦理规范，让人工智能为创造人类共同价值服务。

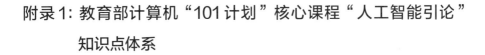

附　录

附录1：教育部计算机"101计划"核心课程"人工智能引论"知识点体系

2021年底，教育部在北京大学启动实施的计算机领域本科教育教学改革试点工作计划（"101计划"）。"101计划"启动以来，教育部计算机科学类基础学科拔尖学生培养基地建设高校中15所高校的约40位教师参与了核心课程"人工智能引论"建设。参与建设的教师认为人工智能具有多学科交叉综合、渗透力大和支撑性强、高度复杂等特点，呈现技术属性和社会属性高度融合的特色。

以"厚基础、强交叉、养品行、促应用"为理念，按照"厚算法基础、养伦理意识、匠工具平台、促赋能应用"的培养目标，"人工智能引论"设置了10个模块和63个知识点（含9个进阶知识点）。具体内容见附表1。

附表1 "人工智能引论"模块与知识点

模块	知识点
1. 人工智能基础模型与历史发展	可计算理论、图灵机模型和图灵测试、人工智能主流模型（符号主义、连接主义和行为主义）、国内外人工智能发展重要事件

续表

模块	知识点
2. 知识表达与推理	知识表示方法、命题逻辑与谓词逻辑及其推理方法、知识图谱推理、贝叶斯网络与概率推理、因果推理
3. 搜索探寻与问题求解	搜索基本概念、贪婪最佳优先搜索、启发式搜索（A*搜索）、搜索算法的性能分析、mini-max 搜索、alpha-beta 剪枝搜索和蒙特卡洛树搜索
4. 机器学习	机器学习基本概念、机器学习模型评估与参数估计、线性回归模型、决策树、聚类、特征降维、演化学习和进阶机器学习等
5. 神经网络与深度学习	人工神经网络概述、感知器模型、梯度下降和误差、反向传播算法、卷积神经网络、循环神经网络、注意力机制、网络优化与正则化、进阶深度学习算法等
6. 强化学习	强化学习基本概念、马尔可夫决策过程、贝尔曼方程、基于表格求解法的策略评估与优化、强化学习中探索与利用的平衡、基于近似求解法的策略评估与优化以及基于策略的强化学习
7. 人工智能博弈	博弈论概念与纳什均衡、虚拟遗憾最小化算法、Gale-Shapely 算法、多智能体博弈算法
8. 人工智能伦理与安全	可信公平人工智能、人工智能可解释性、算法攻击与防守
9. 人工智能架构与系统	人工智能工具（算法支撑技术链）、人工智能芯片（GPU、XPU 和类脑芯片等）和人工智能平台（分布式深度学习优化）
10. 人工智能应用	利用人工智能模型和算法来实现自然语言中的机器翻译、视觉理解中的图像分类、机器人中的行为控制和科学计算以及大模型等具体例子

附图1为"人工智能引论"核心课程知识点构成图。

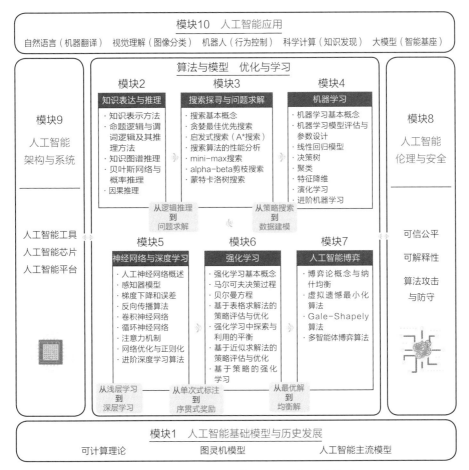

附图1 "人工智能引论"核心课程知识点构成

人工智能10个知识点模块之间相互支撑、互为一体，将算法、模型、系统、应用和伦理规范等有机结合，在算法和模型方面，强调了从逻辑推理到问题求解、从策略搜索到数据建模、从浅层学习到深层学习，从学习结果误差评价到序贯式反馈收益评估，从最优解优化到均衡解博弈。

附录 2：ACM 和 IEEE-CS 制定的新版人工智能知识点

国际计算机学会（ACM）和电气与电子工程师协会计算机分会（IEEE-CS）从 2021 年开始为计算机本科专业制定人工智能领域知识点（详见：https://csed.acm.org/knowledge-areas-intelligent-systems-ai-sigcse-2022-version/），强调了神经网络和表示学习在该领域的重要性。同时，搜索在人工智能中的关键作用仍倍受重视，但符号主义人工智能方法略有减少，以增加对神经网络等内容的关注。人工智能在医学、可持续性、社交媒体等领域的实际应用也日益受到重视。此外，人工智能技术对社会的广泛影响，包括伦理、公平、可信和可解释性等问题，也需要引起关注。考虑到人工智能与其他知识领域的广泛联系，每位计算机科学专业的学生都应该明确目标，培养基本的人工智能素养和批判性思维。

附表 2 给出了 ACM 和 IEEE-CS 联合工作组正在制定的人工智能知识点，包括基本问题、基本搜索策略、基础知识表示和推理、基础机器学习、应用和社会影响、高级搜索、高级表示和推理、不确定下的推理、智能体、自然语言处理、高级机器学习、机器人、感知和计算机视觉等 13 个模块。

附表 2 ACM 和 IEEE-CS 制定的人工智能领域知识点

知识模块	知识点
基本问题 Fundamental Issues	人工智能问题概述，以及最近成功的 AI 应用示例；智能行为定义；图灵测试；理性推理与非理性推理；问题特征；智能体本质；人工智能的哲学问题

续表

知识模块	知识点
基本搜索策略 Basic Search Strategies	问题的状态空间表示；无信息搜索；启发式搜索；搜索算法的空间和时间复杂度；双人博弈
基础知识表示和推理 Basic Knowledge Representation and Reasoning	知识表达类型；概率推理回顾，贝叶斯定理；贝叶斯推理
基础机器学习 Basic Machine Learning	机器学习任务的定义和示例；基于统计的监督学习（朴素贝叶斯和决策树）；机器学习优化（如最小二乘回归）；过拟合问题和正则化；机器学习评估；神经网络基础；表达
应用和社会影响 Applications and Societal Impact	人工智能在广泛问题和不同领域的应用（如医学、可持续发展、社交媒体等）；人工智能的社会影响
高级搜索 Advanced Search	构建搜索树、动态搜索空间、搜索空间的组合爆炸；随机搜索；模拟退火算法；遗传算法；蒙特卡洛树搜索；A*搜索、束搜索（beam search）、最小最大值搜索、alpha-beta 剪枝搜索的实现；期望最大搜索（MDP 求解）和机会点
高级表示和推理 Advanced Representation and Reasoning	命题逻辑和谓词逻辑的回顾；归纳和定理证明（仅命题逻辑）；知识表示问题；描述逻辑；本体工程；非单调推理（例如，非经典逻辑、默认推理）；论证；关于行动和变化的推理（例如，情况和事件演算）；时空推理；基于规则的专家系统；语义网络；基于模型和案例的推理；规划

续表

知识模块	知识点
不确定下的推理 Reasoning Under Uncertainty	基本概率论回顾；随机变量和概率分布；概率公理；概率推理；贝叶斯法则；条件独立；知识表示；精确推理及其复杂度；随机抽样（蒙特卡洛）方法（如吉布斯采样）；马尔可夫网络；关系概率模型；隐马尔可夫模型；决策理论
智能体 Agents	智能体定义；智能体结构（如反应、分层和认知）；智能体理论；理性与博弈论；智能体决策理论；马尔可夫决策过程；软件智能体、个人助理与信息访问；学习智能体；多智能体系统
自然语言处理 Natural Language Processing	确定性语法和随机语法；解析算法；上下文无关文法（CFG）和图表解析器（例如CYK）；概率CFG和加权CYK；表示意义和语义；基于语料库的方法；N元文法和隐马尔可夫链；平滑和回退；应用示例：词性标注和语言形态学；信息检索；词频和逆文档频率（TF&IDF）；查准率和查全率；信息抽取；语言翻译；文本分类
高级机器学习 Advanced Machine Learning	各类机器学习任务的定义和示例；基于统计的广泛学习，参数估计（最大似然）；归纳逻辑程序设计（ILP）；监督学习；决策树；简单的神经网络/多层感知器；支持向量机；集成学习；最近邻算法；深度学习；无监督学习和聚类；半监督学习；学习图模型；性能评估（例如交叉验证和接受者操作特征[ROC]曲线）；学习理论；过拟合的问题，维度灾难问题；强化学习；机器学习算法在数据挖掘中的应用

续表

知识模块	知识点
机器人 Robotics	机器人问题与进展概述；机器人系统（包括传感器和传感器处理等）；机器人控制架构（例如，协商式与反应式控制和Braitenberg车）；世界空间建模和世界空间模型；传感和控制中的固有不确定性；轨迹规划和环境地图；解释传感器数据中的不确定性；定位；导航和控制；运动规划；多机器人协作
感知和计算机视觉 Perception and Computer Vision	计算机视觉；图像采集、表示、处理和图像特点；形状表示、对象识别和分割；运动分析；音频和语音识别；识别中的模块化；模式识别方法

附录3：联合国教科文组织发布的《K-12 AI 课程：政府认可的 AI 课程图谱》

联合国教科文组织一直重视K-12人工智能教育，认为所有公民都需要具备一定程度的人工智能能力，包括具备"人工智能素养"中的知识、理解、技能和价值观，因为这已成为21世纪的基本语法。联合国教科文组织发布的《K-12 AI 课程：政府认可的 AI 课程图谱》见附表3。

附表3　联合国教科文组织《K-12 AI 课程：政府认可的 AI 课程图谱》

类别	知识点	课程中有关人工智能素养的描述
人工智能基础	算法和编程	与数据所起的作用一样，算法和编程是参与人工智能的技术基础
	数据素养	大多数人工智能应用程序运行在"大数据"上。掌握数据全生命周期中收集、清洗、标记、分析和结果报告等内容是使用和/或开发人工智能的技术基础之一。了解数据及其功能也可帮助学生理解人工智能所面临若干道德和部署挑战的背后原因及其在社会中的作用
	情境式解决问题	人工智能通常被认为是应对商业或社会挑战的潜在解决方案，为此需要形成针对具体任务和场景的情境式解决问题框架，包括设计思维和基于项目的学习

续表

类别	知识点	课程中有关人工智能素养的描述
伦理与社会影响	人工智能伦理	无论技术专业和职业背景如何，未来社会的学生都将在个人和职业生活中应用人工智能。对于每个人而言，理解人工智能的伦理挑战非常重要，需要了解人工智能伦理，形成可解释、可信和公平等概念；知晓在不道德或非法使用人工智能的情况下（例如包含有害偏见或侵犯隐私权）如何采取防范措施或补救办法
	人工智能社会影响	人工智能的社会影响涵盖法律框架和劳动力转型等方面。要理解人工智能时代的职业代替、法律框架变化和人工智能治理等趋势
	人工智能在信息技术以外的应用	人工智能在计算机科学之外有着广泛的应用，诸如艺术、音乐、社会研究、科学和健康等
理解、使用和开发人工智能	理解和使用人工智能理论	理解人工智能基本理论（如模式定义与识别、机器学习模型），让学生能够使用现有人工智能算法（如训练分类器），了解人工智能中的算法分类（如监督学习、非监督学习、强化学习、深度学习和神经网络等）
	理解和使用人工智能技术	人工智能技术通常可以完成人类所需的应用，可以"作为服务"提供（如自然语言处理和计算机视觉等）。知晓应用现有的人工智能技术来完成任务或项目，研究创建这些技术的过程

续表

类别	知识点	课程中有关人工智能素养的描述
理解、使用和开发人工智能	开发人工智能技术	开发人工智能技术涉及创建新的人工智能应用程序，这些应用程序可能会解决社会挑战或提供新型服务。这是一个专业领域，需要了解编程、数学（尤其是统计学）和数据科学等方面的一系列复杂技术和技能

2022年2月，联合国教科文组织发布了《K-12 AI课程：政府认可的AI课程图谱》（K-12 AI Curricula: A Mapping of Government-Endorsed AI Curricula），这是关于K-12人工智能课程全球状况的第一份报告。在这份报告中，联合国教科文组织对K-12的人工智能教育提出了9个知识点领域，分别是算法和编程、数据素养、情境式解决问题、人工智能伦理、人工智能社会影响、人工智能在信息技术以外的应用、理解和使用人工智能理论、理解和使用人工智能技术、开发人工智能技术。

English Version

Part I

Challenges Posed to Higher Education by the Advancement of Artificial Intelligence

Sailing from Dartmouth to the New Generation Artificial Intelligence

In August 1955, four scholars submitted a proposal to the Rockefeller Foundation of the United States entitled "A Proposal for the Dartmouth Summer Research Project on Artificial Intelligence"[①]. The four scholars were: John McCarthy, an assistant professor at the Department of Mathematics, Dartmouth College and the recipient of the 1971 Turing Award; Marvin Lee Minsky, a young researcher at the Department of Mathematics and the Department of Neurology, Harvard University, who later received the 1969 Turing Award; Claude Shannon, a mathematician at Bell Labs renowned as father of information theory; and Nathaniel Rochester, head of information research at IBM and chief designer of the 701, IBM's first generation of general-purpose computers.

[①] *John McCarthy, Marvin Minsky, Nathaniel Rochester, Claude Shannon: A proposal for the Dartmouth summer research project on artificial intelligence, 1955*

The proposal hoped that the Rockefeller Foundation would fund a symposium on AI that would be held in the summer of 1956 at Dartmouth College. The term "artificial intelligence" (AI) was used for the first time in this proposal.

Approximately 40 scholars participated in this AI symposium at Dartmouth College from 18 June to 17 August 1956, discussing topics such as automatic computation to simulate high-level functions of the human brain, the use of a common language for computer programming to mimic the reasoning of the human brain, the interconnection of neurons to form concepts, measures of computational complexity, algorithmic self-improvement, the abstraction ability of algorithms, randomness and creativity. This conference marks the inception of humanity's journey towards AI.

Over the course of the following 70 years, connectionism (represented by neural networks), symbolism (exemplified by knowledge engineering), and behaviorism (illustrated by control theory) have intertwined in their progress, nurturing mutual advancement. At the beginning of the 21st century, AI methodologies, epitomized by "deep learning", achieved remarkable progress in fields such as computer vision, speech recognition, and gaming.

China's AI development is keeping pace with evolving times. The "New Generation Artificial Intelligence Development Plan" was unveiled by the State Council in 2017, outlining five key forms of intelligence: big data-driven cognitive learning, cross-media collaborative processing, swarm integrated intelligence, human–machine collaboration-strengthened intelligence, and autonomous intelligent systems. This plan highlighted the emerging features of the new-generation AI, including deep learning, interdisciplinary integration, human–computer collaboration, openness to collective intelligence, and autonomous operation. This pivotal moment signifies China's transition into a new stage of AI development.

The emergence of the large language model (LLM) computing paradigm in late 2022 drove the construction of a new computational architecture for AI that moves from one model solving one task to one model solving all tasks (all in one). At the heart of this architecture is GenAI, which features powerful content synthesis capabilities that drive breakthroughs in areas such as language generation and conversational AI. The development of GenAI will further popularize and apply AI technology, increasing convenience and innovation for society.

Challenges and Opportunities of Artificial Intelligence for Higher Education

Like transformative inventions such as steam engines, electricity, computers, and the internet throughout history, AI has evolved into a general-purpose technology, rapidly reshaping every aspect of human society. According to "The Future of Jobs Report 2023" published by the World Economic Forum, AI is projected to lead to the displacement of 85 million jobs globally by 2025, while creating 97 million jobs simultaneously, resulting in a net increase of 12 million jobs[①]. The report also identifies artificial intelligence and big data competencies as among the top ten skills crucial for future development.

Recognizing the prevalence and significance of AI applications, UNESCO introduced the "AI Competency Frameworks for School Teachers and Students" in 2023. The frameworks emphasize AI literacy as a fundamental skill for both students and teachers. It underscores the importance of acquiring the knowledge, skills, and attitudes necessary to comprehend and utilize AI safely and effectively in educational contexts and beyond[②]. The shifting landscape of employment needs and skill requirements driven by AI is compelling higher education institutions to optimize and innovate their approaches to talent development.

1. Transformation of Knowledge Production Triggered by Artificial Intelligence

Through intelligent algorithms, especially neural network architectures such as Transformer[③], AI can learn the symbiotic correlations between words from massive corpora and achieve natural language synthesis. GenAI systems such as ChatGPT, built with Transformer as the core, have rapidly developed the ability to

---

① *https://www.weforum.org/publications/the-future-of-jobs-report-2023/*
② *https://www.unesco.org/en/digital-education/ai-future-learning/competency-frameworks*
③ *Vaswani, Ashish, Noam Shazeer, Niki Parmar, Jakob Uszkoreit, Llion Jones, Aidan N. Gomez, Łukasz Kaiser, and Illia Polosukhin: Attention is all you need, Proceedings of Advances in Neural Information Processing Systems, 2017*

synthesize natural language, by learning the correlation between words and sentences in massive data, steadily increasing the model size according to the scaling law and exceeding the Feynman limit to enhance the model's nonlinear mapping ability.

Through its robust natural language processing and generation capabilities, GenAI not only generates personalized content tailored to users' needs and preferences but also processes and comprehends vast corpora. This signifies that knowledge production is no longer entirely dependent on the capabilities and time costs of individuals; rather knowledge can be efficiently and extensively generated through algorithms. This transformation enhances the efficiency and speed of knowledge production, and introduces novel opportunities for the integration, dissemination, and innovation of human knowledge. Consequently, it propels higher education towards a more intelligent and informative trajectory.

2. Transformation of the Higher Education Teaching Mode Triggered by Artificial Intelligence

Many institutions of higher education around the world have been actively exploring AI applications in classroom settings. When applied effectively, AI can empower tools, such as virtual teaching assistants, chatbots, and intelligent assessment systems, which significantly enhance the efficiency and quality of instruction. GenAI educational applications are poised to transform the traditional "teacher–student" dichotomy into a "teacher–machine–student" ternary structure. This shift will foster ubiquitous learning environments, meet personalized needs throughout the learning process, and create collaborative spaces for human–machine interaction.

Implementing a closed-loop collaborative learning mechanism, known as "human-in-the-loop" and developing a theoretical model of computation that integrates data-driven induction, knowledge-guided deduction, and feedback-based epiphany, is essential for the future "teacher–machine–student" integrated learning space.

3. Transformation of the Research Paradigm in Higher Education Triggered by Artificial Intelligence

AI is being fully integrated into sci-tech and engineering research, assisting researchers in generating hypotheses, designing experiments, calculating results, and explaining mechanisms, especially in conducting numerous repetitive verifications and trial-and-error processes under various hypotheses. This human–machine collaborative model greatly accelerates scientific innovation, empowering researchers to open up previously inaccessible knowledge horizons and fields.

LLM interacts with humans through natural language and integrates various applications as plugins, becoming a gateway linking the triadic Cyber–Physical–Human (CPH) space. Additionally, intelligent entities—AI models capable of perceiving their environment, making autonomous decisions, and taking action—are combined with GenAI-based models to form the advanced frontier of AI agents. These AI agents autonomously perform tasks such as information retrieval, human–computer dialog, task execution, and logical reasoning based on content synthesis.

Applying agents to research enhances the process of generating hypotheses, designing experiments, calculating results, and interpreting mechanisms. This integration not only advances human learning capabilities but also broadens the research paradigm in a previously uncharted territory.

4. Potential Negative Impacts of Artificial Intelligence Applications in Higher Education

The emergence of GenAI has made intelligent machines an adjunct to knowledge production, challenging the capacity of individual learners to independently think, judge and learn, as well as their ethical and moral values.

The application of AI in education will also have many negative impacts, such as the marginalization of teachers' status, the "island" of student learning, the fragmentation of knowledge systems, the leakage of private information, discrimination and prejudice, ethical risks, academic integrity and fairness imbalance, and the alienation of educational relationships, security and privacy issues, knowledge blind spots and information cocoons, insufficient content accuracy, weakened students' higher-order thinking, and the digital application divide.

The new generation of AI technology is characterized by deep learning, cross-border integration, human–machine collaboration, openness to collective intelligence, and autonomous control. However, its low explainability, system deviation, data security, data privacy and other issues have also posed unprecedented challenges to society.

"In the era of intelligence, what is higher education for?" Teachers in Chinese institutions of higher education should accurately identify changes, respond scientifically, take initiative to seek changes, and reconstruct the objectives, paths, and support systems of talent cultivation for all majors in the intelligent era so as to cultivate a large number of comprehensive professionals with AI literacy who will help future society move towards harmonious, healthy and sustainable development.

Part II

Objectives and Vision of Artificial Intelligence Literacy Cultivation for College Students

Concept and Constitutive Connotation of Artificial Intelligence Literacy

In Chinese, the word "literacy"(素养) originated from Liu Biao's Biography in Volume 74 of *Hou Han Shu* (*Book of the Later Han*). "Those who have higher levels of literacy can be motivated and bring others together to participate in or support an action or decision if benefits or advantages are shown to them."[①] The term refers to daily personal accomplishment (修养), which in a broader sense encompasses all aspects of moral character, appearance, knowledge and ability.

The Organization for Economic Cooperation and Development (OECD) conducted extensive research between 1997 and 2005 on the definition and selection of literacy, focusing on its theoretical and conceptual foundations. This nine-year study culminated in the definition of literacy as "the ability to use and mobilize psycho-social resources (including skills and attitudes) to meet complex needs in

---

① "越有所素养者,使人示之以利,必持众来" *in Chinese*

specific contexts"[1]; literacy is contemporary, holistic, developmental, and measurable. Literacy, according to this definition, depends on specific contexts, and any abilities and skills that help individuals adapt to society or solve complex problems can be considered literacy. Encompassing the unity of knowledge, abilities, and attitudes, it is developed through specific educational means, and measurable through comprehensible, actionable, and evaluable indicators.

In 2018, UNESCO further defined literacy as "the ability to identify, understand, interpret, create, communicate and compute, using printed and written materials associated with varying contexts".

According to previous definitions, we define literacy as comprehensive performance. In specific situations, when faced with complex and changing problems in life, individuals can utilize and mobilize their knowledge, skills, judgment, and other abilities to observe and understand the world, construct solutions to problems, and test the rationality of their understanding through their actions. This process enhances their competence and contributes to the incremental knowledge of humanity.

In the 2020 version of the Computing Curricula, the Association for Computing Machinery (ACM) and the Institute of Electrical and Electronics Engineers Computer Society (IEEE-CS) adopted the concept of "competence" to describe professional computing

---

[1] *Marilyn Mathieu: The definition and selection of key competencies: Executive summary, 2005*

competence, which consists of three dimensions: knowledge, skills, and dispositions.

As humans enter the era of intelligence, AI literacy is gradually becoming one of the most important literacies for individual survival and development. The concept was first proposed in the 1970s, when the focus was on the literacy composition of AI professionals and technicians[①]. Given the enormous impact of AI on human society, every citizen needs to understand what AI is, what AI can and cannot do, how to use AI responsibly, and how to challenge the use of AI, so that AI can serve individual, group, and public interests.

In recent years, UNESCO has been focusing on global AI education, arguing that all citizens need to be "AI-literate" in terms of knowledge, understanding, skills, and values, which has become the basic grammar of the 21st century.

The Goal of Cultivating College Students' Artificial Intelligence Literacy

As the primary focus of higher education and prospective leaders across various sectors of society, college students garner considerable attention and expectations. Typically, these students are young and energetic, with well-established learning abilities and habits. They

---

① *Agre, Philip E.: What to read: A biased guide to AI literacy for the beginner, MIT Artificial Intelligence Laboratory Working Papers, WP-239, 1972*

exhibit substantial plasticity, creativity, and developmental potential. Given these attributes and the distinct expectations of both the country and society, the connotation of AI literacy for college students in the age of intelligence diverges from that of other social cohorts.

To address the unprecedented challenges posed by AI across all sectors, institutions of higher education must cultivate students with comprehensive capacity to comprehend, utilize and innovate with AI, and adhere to the ethical relationship between humans and machines in the AI era. Augmenting their AI literacy will empower them to evolve and excel within an intelligent society.

Specifically, the AI literacy of college students encompasses four integral components: systematic knowledge, constructive capability, creative value, and humanistic ethics. Systematic knowledge lays the bedrock, constructive capability forms the core, creative value lights the path ahead, and humanistic ethics remains the essence evermore. These four components complement and integrate with each other.

Systematic knowledge: Cognition is an important manifestation of human intelligence, the cornerstone of which is systematic knowledge, which means that learning objects can be understood holistically and classified systematically. The AI has cross-penetration characteristics ranging from smallest to largest. Once systematic knowledge is acquired, one can understand the connotations, boundaries and extensions of AI more clearly.

Systematic knowledge is a relative and dynamic concept. One

of the fundamental tasks of human thinking is to classify all kinds of complicated things, gradually forming systematic and comprehensive thinking, improving general cognition, and reaching consensus.

Constructive capability: AI can help establish a cooperative relationship between humans and machines, taking into account the respective advantages of AI and human intelligence, and working together to achieve the goals of specific tasks. During the human–machine collaboration process, humans can gain more insights from data, determine the optimal solution, and complete tasks in a new auxiliary way that was previously unseen.

Science has long followed a cycle from hypothesis to experiment to theory verification, with its core being the search for explainable principles behind phenomena. In the era of AI, people can develop the ability to solve problems in a closed-loop system with humans in the loop by using AI tools. This involves abstract modeling of problems, generating verifiable hypotheses, designing computable models, interpreting algorithm results, and continuously optimizing solutions through methods like enumeration and simulation based on feedback. The constructive capabilities overcome the difficulty of traditional methods in handeling complex data, driving a shift from "knowledge-based education" to "competence-based education".

Creative value: GenAI can compress the context information of all human corpora and then synthesizes it in a probabilistic way. The advantages of the education model centered on knowledge accumulation disappear due to GenAI's powerful ability to memorize and integrate existing knowledge.

Through the enhancement of subjectivity, the expression of individuality, the amplification of inivative, and the participation of enhanced experiments with AI, incremental human knowledge is produced, forming creative value and thereby becoming "universal intelligence" accumulated and shared by society.

Humanistic ethics: In traditional sci-tech development, a so-called "technology first or proactive approach" model is usually adopted, which prioritizes technological progress and exhibits powerful instrumental rationality, that is, "maximizing efficiency or utility through rigorous logical thinking and sophisticated scientific calculations".

With the development of technologies such as the Internet of Things (IoT) and AI, humanity has fully entered the CPH space, namely, the CPH society, in which the academic ethical discussion is no longer just the relationship between humans or the one between humans and established facts of nature, but rather the one between humans and artificial objects in society, so AI has dual attributes of technology and society.

Therefore, in a human–machine-inclusive society, it is imperative to adhere to the ethical principles of "AI for good" and human-centered alignment. Human values, morals, and laws must be integrated throughout AI products and services, ensuring that AI embodies social attributes. Figure 1 illustrates the connotations, carriers, actions, and strategies for college students' AI literacy cultivation.

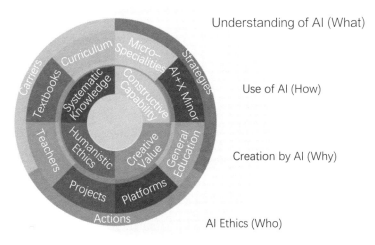

Figure 1 Connotations, carriers, actions, and strategies for college students' AI literacy cultivation

Visions of Artificial Intelligence Literacy Cultivation for College Students

The four major visions for cultivating AI literacy among college students include the formation of AI discursive modes, the ability to solve problems with AI, the value of creating incremental human knowledge, and adherence to human-centered ethical limits. Table 1 shows the specific contents and visions of cultivating AI literacy among college students from the four constituent dimensions of knowledge, capability, value, and ethics.

Table 1 Components, contents, and visions of college students' AI literacy

Components	Contents	Visions
Systematic Knowledge	■ Data and knowledge: Fuel of AI ■ Algorithms and models: Engine of AI ■ Computility and systems: Carrier of AI ■ Interdisciplinary application: Application of AI ■ Trustworthiness and security: Double-edged sword of AI	Formation of AI discursive patterns: relative transformation of AI's "cans" and "can'ts"; dialectical unity of certainty (logic) and uncertainty (probability) in AI; symbiotic synergy between machine intelligence and natural intelligence; and interdisciplinarity and synthesis of "AI+" and a high degree of integration of sci-tech attributes with social attributes
Constructive Capability	■ Abstraction and modeling of problems to solve ■ Decomposition and modularity of solving process ■ Verifiable assumptions about solving methods ■ Interpretation and feedback on solutions ■ Ability to solve problems using GenAI	Problem-solving skills with AI: cultivation of the computational thinking of design and construction; integration of machine intelligence induction and human intelligence intuition; shaping problem-solving ability through the mechanism of human–computer collaboration; and transformation from "knowledge-based education" to "competence-based education"

Components	Contents	Visions
Creative Value	■ Content reconstruction under goal-directed dialog ■ Enhancement of cognitive subjectivity in teacher–machine–student interactions ■ Autonomous incorporation of personalized learning experience ■ Practical dynamic experience in problem solving ■ Selective introspection in overcoming dependence on intelligent tools	Create the value of incremental human knowledge: enhancing personalization, subjectivity and mobility with the assistance of AI; creating value through content reconstruction and synthesis, practical exploration, interactive cognition, etc.; and transformation from "knowledge learning, ability shaping" to "value creation"
Humanistic Ethics	■ Awareness of data security and privacy protection ■ Vigilance of algorithmic bias and model illusion ■ Alignment of "AI for good" and human-centered AI ■ AI & All concept of human–machine symbiosis and inclusion ■ Pursuit of universal sharing of accumulated human knowledge	Adherence to the human-centered ethical bottom line: knowing the potential hazards of AI vulnerability in terms of data, algorithms, models and applications; understanding the alignment model of "AI for good" and human-centered AI; and establishing AI & All concept of human–computer coexistence and pervasive intelligence

Systematic knowledge encompasses data and knowledge, algorithms and models, computility and systems, and interdisciplinary applications, which are the fuel, engine, carrier, and application of AI. It also includes knowledge of trustworthiness and security necessary to face the double-edged sword of AI. The mastery of systematic knowledge is essential for developing a comprehensive and systematic mindset to address complex problems.

Figure 2 depicts the systematic knowledge graph for the core course Introduction to Artificial Intelligence, one of the core courses of the 101 Plan for Computer Education implemented by the Ministry

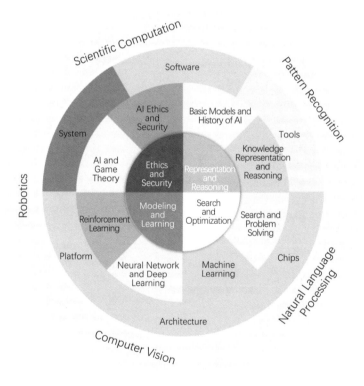

Figure 2 Systematic knowledge of Introduction to Artificial Intelligence, one of the core courses of the 101 Plan for Computer Education implemented by the Ministry of Education (China)

of Education (China) (see Appendix 1). This course is based on the principles of "solid foundation, strong interdisciplinary integration, character cultivation, and application enhancement".

Constructive capability encompasses the abstraction and modeling of problems to solve, the decomposition and modularity of solving process, verifiable assumptions about solving methods, interpretation and feedback on solutions, as well as the ability to solve problems via GenAI, under the human–machine collaborative model.

Creative value includes models such as content reconstruction under goal-directed dialog, enhancement of cognitive subjectivity in teacher–computer–student interactions, autonomous incorporation of personalized learning experiences, practical dynamic experience in problem solving, and selective introspection in overcoming dependence on intelligent tools.

Humanistic ethics include awareness of data security and privacy protection, vigilance of algorithmic bias and model illusion, alignment of "AI for good" and human-centered AI & All concept of human–machine symbiosis and inclusion, and the pursuit of universal sharing of accumulative human knowledge.

Part III

Carriers, Actions, and Strategies for Artificial Intelligence Literacy Cultivation for College Students

Carriers for Artificial Intelligence Literacy Cultivation for College Students

1. Curriculum: Categorized and Graded Courses to Meet Diverse Artificial Intelligence Learning Needs

In higher education, one of the most important carriers for the cultivation of AI literacy among college students is a series of courses, i.e. curricula. According to the requirements of college students of different majors and levels, institutions of higher education can set up multiple categories of courses, such as introductory courses[①], specialized courses, interdisciplinary courses, and micro-courses, to meet the needs of different types of students for AI knowledge. Table 2 presents the different categories of AI courses and the knowledge contents covered.

Table 2 Description of the system and contents of multi-category AI curriculum in higher education institutions

Curriculum Category	Curriculum Contents
Introductory	Introducing the history, definition, branches, applications, and cutting-edge developments of AI; discussing the ethical issues, privacy protection, and security challenges of AI; exploring the impact of AI on the future society, including employment, education, healthcare, etc.

---

① *Often for general education*

Continued

Curriculum Category	Curriculum Contents
Specialized	Instructing different aspects of perception (e.g. speech recognition, natural language understanding, computer vision), problem solving (e.g. search and planning), action (e.g. robotics), and architectures that support task completion (e.g. agents and multi-agent systems)
Interdisciplinary AI + X	Introducing the basic knowledge of AI and the ways to solve problems in different disciplines with AI, such as AI-empowered economics, law, and art; clarifying the paradigm revolutions generated by AI in different disciplines by solving problems in different disciplinary scenarios
Micro	Providing personalized learning by focusing on knowledge modules and introducing the contents of different knowledge modules of AI related to specific problems

2. Textbooks: Utilization of Digital Teaching Resources to Facilitate the Dissemination of Knowledge

As a new form of teaching materials for the deep integration of digital technologies and education teaching, digital textbooks have the advantages of high integration, strong interactivity and obvious structuring, which provide higher quality and rich digital resources for education and teaching in higher education institutions, and are of great significance in accelerating the digital transformation of education and the construction of a leading country in education. Digital textbooks are not only digital versions of traditional paper textbooks, but also a new form of teaching materials that organically integrate teaching contents, rich media teaching resources, learning tools and technology platforms, which provides the possibility of large-scale "individualized instruction"[①].

While strengthening the advantages of audio and visual materials in transmitting information and greatly expanding the reading boundaries of readers, digital textbooks can also effectively garner readers' attention, enhance learning concentration, improve reading experience, and provide fuel for students' individualized learning and teachers' creative teaching.

Teaching resources such as videos, audios, practical training cases and lecture slides corresponding to AI teaching materials can be organized according to knowledge concepts, thus developing a structure of "knowledge map–ability map". This opens a wide world for "students to learn more actively and teachers to instruct more creatively".

3. Teachers: Multiple Cross-border Integration to Promote the Construction of Teacher Teams

Teachers are the cornerstone and pivotal force driving digital intelligence in education, catalyzing the profound integration of AI technology into teaching and learning. It is essential to foster collaboration and communication among teachers through diverse and interdisciplinary integration.

Attracting interdisciplinary teachers to join and build an

[①] "因材施教" *in Chinese*

open AI teacher team. As a multi-dimensional field, AI intersects with computer science, mathematics, neuroscience, psychology, and beyond. In order to effectively nurture students' AI literacy, educational institutions should actively recruit teachers from various academic domains to contribute to the development of AI-focused curricula.

On the one hand, educators specializing in computer science, automation, and electronic information should be encouraged to transition into AI instruction. Institutions can facilitate this transition by providing necessary training and support, enabling them to swiftly acquire cutting-edge knowledge and pedagogical techniques in AI. Furthermore, these educators should be guided to seamlessly integrate AI concepts into their respective disciplines, fostering the creation of interdisciplinary courses that are both innovative and impactful.

On the other hand, it is equally crucial to invite exceptional teachers from diverse fields, such as mathematics, physics, biology, medicine, economics, management, sociology, and art and design to contribute to AI education. Their unique perspectives and expertise can enrich AI general education and interdisciplinary instruction, providing students with a comprehensive understanding of AI implications across various domains.

Strengthening academia–enterprise cooperation and integrating industry expertise into teaching are pivotal steps in advancing AI education. Given its close ties to industrial applications, leveraging collaboration with enterprises and involving experts and engineers in teaching can significantly enhance the relevance and practicality of AI course contents.

Institutions of higher education can partner with leading AI enterprises, fostering joint laboratories and practical teaching bases. By doing so, institutions create immersive learning environments mirroring real-world industrial settings. Moreover, enterprises can contribute by dispatching outstanding experts and engineers to work as part-time professors, delivering specialized courses and practical instruction.

Guiding young teachers to join AI curriculum groups. As burgeoning talents within academia, young educators bring fresh perspectives and vigor to AI education. Institutions of higher education should actively encourage and support promising young teachers with a professional background and interest in AI to join dedicated teaching teams. Through mentorship and professional development initiatives, these educators can thrive in their roles, enriching the educational landscape with their insights and dedication.

Actions for Artificial Intelligence Literacy Cultivation for College Students

1. Projects: Diversified Design to Promote the Integration of Theory and Practice

Institutions of higher education should establish a diversified and open AI training project system, focusing on cutting-edge technologies and application scenarios. This approach encourages interdisciplinary collaboration and academia–enterprise cooperation, guiding students to enhance their knowledge, skills, and values, with a particular emphasis on ethical awareness. By participating in projects such as ethical assessments, interdisciplinary innovation, open-source contributions, industry–academia research collaborations, and social services, students can gain a profound understanding of AI's social impact. They learn to balance technological innovation with ethical standards, contribute value through open collaboration, and assume responsibilities in serving society. Only by closely integrating knowledge learning with practical application can institutions of higher education truly cultivate new-generation AI leaders who possess patriotism, a global vision, innovative capabilities, and ethical literacy.

Institutions of higher education can establish practical projects in areas such as AI ethics, interdisciplinary AI + X, AI open-source ecosystems, AI industry–academia research integration, and AI social services. These projects aim to consolidate the practical training capabilities of college students from various perspectives.

2. Platforms: Building Digital Platforms to Enhance Practical Abilities

Comprehensive digitization of teaching resources. The digitalization of education represents a crucial breakthrough for pioneering new avenues of educational development and creating new advantages in educational growth. It effectively supports personalized learning, lifelong learning, expansion of high-quality educational resources, and modernization of education. By systematically organizing teaching resources such as course contents, instructional micro-videos, and practical training exercises corresponding to each knowledge point, institutions of higher education construct digital teaching resources. This innovation in knowledge carriers drives supply-side structural reform of educational resources, enabling students to learn in a personalized manner and teachers to instruct creatively.

Generative foundation-model-supported learning platforms. GenAI exhibits strong content synthesis capabilities, driving groundbreaking advancements in fields like language generation and conversational AI, demonstrating the significant impact of technological transformation on education. Tailoring to the characteristics of different disciplines, institutions of higher education can develop LLMs for vertical educational sectors through processes such as foundation model construction, model training, model optimization, edge intelligence empowerment, and application iteration. These models support education and teaching, academic research, and educational governance, highlighting the foundational, pioneering, and comprehensive role of education. This support extends from backing educational instruction to academic research, and further to empowering intelligent upgrades across various industries.

Strategies for Artificial Intelligence Literacy Cultivation for College Students

1. General Education in Artificial Intelligence: The Cornerstone of Literacy Cultivation

In educational instruction, the focus is shifting from educators as leaders to learner-centered approaches, and professional development is transitioning towards a coherent development encompassing general, specialized, and interdisciplinary aspects. In the era of intelligence, institutions of higher education should incorporate AI as an essential component of general education.

AI courses should be geared to all college students, disseminating basic AI concepts, introducing the status quo, analyzing future trends, guiding college students to contemplate the societal changes brought about by AI, and cultivating humanistic sentiments and a sense of mission.

2. AI + X Minor: Promoting Interdisciplinary Talent Cultivation

To cultivate high-level AI compound talents with interdisciplinary integration and innovation capabilities, institutions of higher education should actively explore the AI + X vertical interdisciplinary talent cultivation model. Building upon undergraduates majoring in AI or computer science with a focus on AI, institutions of higher education should guide and support them in interdisciplinary learning during their graduate studies, expanding their knowledge horizons and enhancing their innovative practical abilities.

Specifically, methods such as **comprehensive AI + X undergraduate-to-graduate programs, interdisciplinary courses, interdisciplinary research training, a cross-supervisor system for graduation projects, flexible credit recognition mechanisms, industry–academia research–application collaborative cultivation, and international exchange and cooperation can be adopted to interdisciplinary talent cultivation.** By emphasizing the cultivation of interdisciplinary literacy in undergraduate education in AI, institutions of higher education can lay a solid foundation for students to deepen their interdisciplinary learning during graduate studies. Additionally, institutions of higher education should provide students with diverse interdisciplinary learning resources and practical opportunities to stimulate their interest and potential in exploring interdisciplinary fields. Institutions of higher education should also explore effective models of cultivating innovative talents with AI + X compound skills according to their own characteristics and advantages, steadily supplying high-level talents for the progress of intelligent societies.

3. Micro-specialties: Interdisciplinary Collaboration to Promote AI + X Talent Cultivation

A micro-specialty consists of a set of micro-courses in specific fields. Micro-courses help learners acquire relatively independent and complete unit knowledge in a short period. Therefore, a micro-specialty provides professionals with potential opportunities to expand their abilities without leaving current positions, and keep their professional knowledge up-to-date.

Micro-specialties can break down barriers between departments and disciplines, leverage the collaboration between the government, institutions and enterprises, bring together top scholars and industrial experts to jointly offer courses, and realize interdisciplinary teaching and management across colleges, disciplines, and majors. This enables students from non-computer science majors and non-AI majors to acquire and understand basic AI knowledge more flexibly and efficiently, master fundamental AI knowledge, and enhance practical application abilities, thus promoting paradigm shifts in future disciplinary studies.

Micro-specialties can be implemented in various flexible and diverse ways, fully leveraging the characteristics of interdisciplinary integration, and cultivating students' AI literacy and practical innovation abilities. Micro-specialties emphasize the proactive integration of traditional disciplines with AI. Specific measures may include **intercollegiate alliance cooperation in education, cross-departmental collaborative course development, hybrid online and offline teaching, and industry–academia collaborative education.**

In summary, the establishment of AI micro-specialties in institutions of higher education should embody the concepts of interdisciplinary integration and open collaboration, break down barriers between disciplines, promote inter-institution, academia–enterprise, and international cooperation, and build and share high-quality educational resources together. Moreover, innovative educational teaching models should be created, which combine online and offline methods, focus on cultivating practical innovation abilities, and adopt multiple measures to improve the quality of talent cultivation. Only through continuous exploration and innovation can micro-specialties in AI become an effective way to cultivate application-oriented talents and better serve economic and social development.

Conclusion

Education is the process of transforming human beings from natural to social beings. In this process, human beings need to maintain continuous interaction with the environment in order to survive and develop, which highlights knowledge, skills, values and ethics involved in literacy.

In *Shuowen Jiezi (Origin of Chinese Characters)*, the words "teach"[①] and "educate"[②] mean respectively "to demonstrate for the lower class to imitate" and "to raise a moral child". The AI era has had a profound impact on educational methods and means.LLMs represented by ChatGPT has exerted a huge impact on education, involving artifacts such as machines and systems in the teaching and learning process and being one of them, instead of just carriers and expression tools of knowledge.

The primary objective of education should always be the cultivation of overall capabilities for independent thought and judgment,

---

① "教" *in Chinese*
② "育" *in Chinese*

rather than the acquisition of specific knowledge. To achieve this, we must adopt a scientific and rational approach to grasp the intrinsic characteristics and potential of technology, engage in comprehensive evaluation and philosophical reflection on technological advancements, and remain vigilant and observant of technological breakthroughs. This will enable a deeper understanding and exploration of the potential value that technology can bring to the field of education.

Whether AI can achieve ethical legitimacy does not depend on AI itself, but rather on whether human beings, as well as AI, are truly directed to the purpose of legitimacy. Facing the future and uncertainty, we should shift our attitude from knowledge-based to competence-based education, abide by the ethical norms of AI development, and put AI at the service of creating shared human values.

Appendices

Appendix 1 Syllabus of Introduction to Artificial Intelligence, one of the core courses of the 101 Plan for Computer Education implemented by the Ministry of Education (China)

The Ministry of Education (China) launched a pilot work plan for undergraduate education reform in the field of computer science (now referred to as the 101 Plan) in December 2021[①]. Since the launch of the 101 Plan, about 40 teachers from 15 universities have participated in developing the core course Introduction to Artificial Intelligence. Teachers involved in the development believe that artificial intelligence possesses characteristics such as interdisciplinary integration, strong penetration and support, and high complexity, demonstrating a high degree of fusion between technical and social attributes.

---

① *Within two years of pilot implementation, the 101 Plan contributed to 12 core courses, 31 core textbooks, and a practice platform covering over 400 projects, involving more than 1,500 teachers in collective teaching preparation and research activities and benefiting 20,000-odd students. See more at http://en.moe.gov.cn/news/press_releases/202404/t20240422_1127028.html*

Guided by the principles of "solid foundation, strong interdisciplinary integration, character cultivation, and application enhancement", and following the training objectives of "fundamental theory understanding, algorithmic methods utilization, ethical and secured challenges and opportunities investigation, practical AI system application", the course Introduction to Artificial Intelligence comprises 10 modules and 63 knowledge points (including 9 advanced knowledge points). The specific content is as follows:

- The understanding of fundamental theory in AI. Students are required to know the basic mechanism of how AI is committed to realizing machine-borne intelligence. Topics include knowledge representation, Symbolistic inference, learning optimization and advanced topics.
- The utilization of algorithmic methods in AI. Students are required to know mainstream AI models such as Symbolism, Connectionism, Behaviorism as well as game theory.
- The investigation of the ethical and secured challenges and opportunities posed by AI. Students are required to know the responsible design and trustworthy deployment of AI systems for daily life, which perform exactly as intended to guarantee the security, fairness and reliability of AI systems.
- The practical application of AI systems. Students are required to build up different AI applications from scratch or implement various existing AI tools, chips and frameworks. Programming

projects include machine translation, computer vision, robotics and scientific computation, etc.

Figure A.1 shows the composition of the knowledge modules in the Introduction to Artificial Intelligence course and Table A.1 lists the topics covered therein.

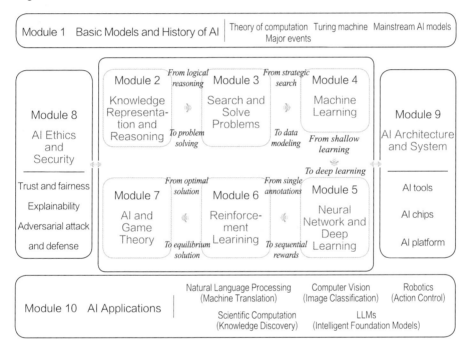

Figure A.1　Composition of the knowledge modules in Introduction to Artificial Intelligence

Table A.1　Topics covered in Introduction to Artificial Intelligence

Modules	Topics
1. Basic Models and History of AI	Theory of computation; Turing machine; mainstream AI (Symbolism, Connectionism, and Behaviorism); major events in AI development in China and the world

Continued

Modules	Topics
2. Knowledge Representation and Reasoning	Knowledge representation; propositional and predicate logic and inference; reasoning over knowledge graph; Bayes network and probabilistic inference; causal inference
3. Search and Problem Solving	Greedy best-first search; A* search; search performance; mini-max search; alpha-beta pruning; Monte-Carlo tree search
4. Machine Learning	Machine learning basics; evaluation and parameter estimation of learning models; linear regression model; decision tree; clustering; feature dimensionality reduction; evolutionary learning; advanced machine learning* (boosting, topic model, non-matrix factorization, hidden Markov model, probabilistic graphical models)
5. Neural Network and Deep Learning	Introduction to artificial neural network; perceptron; gradient descent and error; back propagation; convolutional neural network; recurrent neural network; attention mechanism; optimization and regularization; advanced deep learning* (generative adversarial learning, graph neural network)
6. Reinforcement Learning	Introduction to reinforcement learning; Markov decision process; Bellman equation; tabular solution methods for policy evaluation and improvement; tradeoff between exploration and exploitation; approximate solution methods for policy evaluation and improvement; policy-based reinforcement learning
7. AI and Game Theory	Basics of game theory and Nash equilibrium; counterfactual regret minimization; Gale-Shapely; multi-agent game theory*

Continued

Modules	Topics
8. AI Ethics and Security	Trustworthy and fair AI; explainable AI; adversarial attack and defense in AI
9. AI Architecture and System	AI tools (algorithm-supported technology chain); AI chips (GPU, XPU, and neuromorphic chips); AI platform (distributed deep machine learning)
10. AI Applications	Machine translation; image classification; robotics; AI for science; LLMs

Appendix 2　The new version of artificial intelligence knowledge areas formulated by the ACM and IEEE-CS

The ACM (Association for Computing Machinery) and IEEE-CS (IEEE Computer Society) began to formulate AI knowledge areas (KA) for computer undergraduate majors in 2021 (see https://csed.acm.org/knowledge-areas-intelligent-systems-ai-sigcse-2022-version/).

KA has changed in the following ways compared with CS 2013:

● The name has changed from "Intelligent Systems" to "Artificial Intelligence", reflecting the most common term used for these topics within the field and the more widespread use of the term outside of the field.

● There is increased emphasis on neural networks and representation learning, reflecting the recent advances in the field. Search is still emphasized due to its key role throughout AI, but there is a slight reduction in symbolic methods in favor of understanding subsymbolic methods and learned representations.

● There is an increased emphasis on practical applications of AI, including a variety of areas (e.g. medicine, sustainability, and social media)

● The curriculum reflects the importance of understanding and assessing the broader societal impacts and implications of AI methods and applications, including issues in AI ethics, fairness, trust, and explainability.

● The AI knowledge area includes connections to data science through cross-connections with data management.

● There are explicit goals to develop basic AI literacy and critical thinking in every computer science student, given the breadth of interconnections between AI and other knowledge areas in practice.

Table A.2 shows the AI knowledge areas formulated by the ACM and IEEE-CS joint working group, which includes 13 modules, covering fundamental issues, basic search strategies, basic knowledge representation and reasoning, basic machine learning, applications and societal impact, advanced search, advanced representation and reasoning, reasoning under uncertainty, agents, natural language processing, advanced machine learning, robotics, perception and computer vision.

Table A.2　AI-SIGCSE 2022 version artificial intelligence knowledge areas

Knowledge Areas	Topics
Fundamental Issues	Overview of AI problems, examples of successful recent AI applications; What is intelligent behavior?; the Turing test; rational versus non-rational reasoning; problem characteristics; nature of agents; philosophical issues
Basic Search Strategies	State space representation of a problem; uninformed search; heuristic search; space and time complexities of search algorithms; two-player games

Continued

Knowledge Areas	Topics
Basic Knowledge Representation and Reasoning	Types of representations; review of probabilistic reasoning, Bayes theorem; Bayesian reasoning
Basic Machine Learning	Definition and examples of a broad variety of machine learning tasks; simple statistical-based supervised learning such as Naive Bayes, decision trees; formulation of simple machine learning as an optimization problem, such as least squares regression; the over-fitting problem and regularization; machine learning evaluation; basic neural networks; representations
Applications and Societal Impact	Applications of AI to a broad set of problems and diverse fields, such as medicine, sustainability, and social media; societal impact of AI
Advanced Search	Constructing search trees, dynamic search space, combinatorial explosion of search space; stochastic search; simulated annealing; genetic algorithms; Monte-Carlo tree search; implementation of A* search, beam search; mini-max search, alpha-beta pruning; expectimax search (MDP-solving) and chance nodes

Continued

Knowledge Areas	Topics
Advanced Representation and Reasoning	Review of propositional and predicate logic; resolution and theorem proving (propositional logic only); knowledge representation issues; description logics; ontology engineering; non-monotonic reasoning (e.g. non-classical logics, default reasoning); argumentation; reasoning about action and change (e.g. situation and event calculus); temporal and spatial reasoning; rule-based expert systems; semantic networks; model-based and case-based reasoning; planning
Reasoning Under Uncertainty	Review of basic probability; random variables and probability distributions; axioms of probability; probabilistic inference; Bayes' rule; conditional independence; knowledge representations; exact inference and its complexity; randomized sampling (Monte-Carlo) methods (e.g. Gibbs sampling); Markov networks; relational probability models; hidden Markov models; decision theory
Agents	Definitions of agents; agent architectures (e.g. reactive, layered, cognitive); agent theory; rationality, game theory; decision-theoretic agents; Markov decision processes (MDP); software agents, personal assistants, and information access; learning agents; multi-agent systems
Natural Language Processing	Deterministic and stochastic grammars; parsing algorithms; CFGs and chart parsers (e.g. CYK); probabilistic CFGs and weighted CYK; representing meaning/semantics; corpus-based methods; N-grams and HMMs; smoothing and backoff; examples of use: POS tagging and morphology; information retrieval; TF & IDF; precision and recall; information extraction; language translation; text classification, categorization

Continued

Knowledge Areas	Topics
Advanced Machine Learning	Definition and examples of broad variety of machine learning tasks; general statistical-based learning, parameter estimation (maximum likelihood); inductive logic programming (ILP); supervised learning; learning decision trees; learning simple neural networks/multi-layer perceptrons; support vector machines (SVMs); ensembles; nearest-neighbor algorithms; deep learning; unsupervised learning and clustering; semi-supervised learning; learning graphical models; performance evaluation (such as cross-validation, area under ROC curve); learning theory; the problem of overfitting, the curse of dimensionality; reinforcement learning; application of machine learning algorithms to data mining
Robotics	Overview: problems and progress; state-of-the-art robot systems, including their sensors and an overview of their sensor processing; robot control architectures, e.g. deliberative vs. reactive control and Braitenberg vehicles; world modeling and world models; inherent uncertainty in sensing and in control; configuration space and environmental maps; interpreting uncertain sensor data; localizing and mapping; navigation and control; motion planning; multiple-robot coordination
Perception and Computer Vision	Computer vision; image acquisition, representation, processing and properties; shape representation, object recognition, and segmentation; motion analysis; audio and speech recognition; modularity in recognition; approaches to pattern recognition

Appendix 3 UNESCO K-12 AI curricula: A mapping of government-endorsed AI curricula

UNESCO has consistently emphasized the importance of K-12 education, asserting that all citizens need to possess a certain degree of AI capability. This includes knowledge, understanding, skills, and values encompassed within "AI literacy", which is now considered the fundamental grammar of the 21st century. Table A.3 presents the UNESCO AI curricula.

Table A.3 UNESCO K-12 AI curriculum areas

Category	Topic Areas	Competency and Curriculum Considerations
AI Foundations	Algorithms and programming	Together with data literacy, algorithms and programming can be viewed as the basis of technical engagement with AI
	Data literacy	A majority of AI applications run on "big data". Managing the data cycle from collection to cleaning, labeling, analysis and reporting forms one of the foundations for technical engagement with using and/or developing AI. An understanding of data and its functions can also help students understand the causes of some of the ethical and logistical challenges with AI and its role in society

Continued

Category	Topic Areas	Competency and Curriculum Considerations
AI Foundations	Contextual problem-solving	AI is often framed as a potential solution to business-related or social challenges. Engaging at this level requires a framework for problem-solving in context, encompassing things like design thinking and project-based learning
Ethics and Social Impact	The ethics of AI	Regardless of technical expertise, students in future societies will engage with AI in their personal and professional lives—many do so from a young age already. It will be important for every citizen to understand the ethical challenges of AI; what is meant by "ethical AI"; concepts such as transparent, auditable, and fair use of AI; and the avenues for redress in case of unethical or illegal use of AI, e.g. that which contains harmful bias or violates privacy rights
	The social or societal implications of AI	The social impacts of AI range from requiring adjustments to legal frameworks for liability, to inspiring transformations of the workforce. Survey respondents were asked about the extent to which their curricula targeted these issues. Trends such as workforce displacement, changes to legal frameworks, and the creation of new governance mechanisms were given as examples

Continued

Category	Topic Areas	Competency and Curriculum Considerations
Ethics and social impact	Applications of AI to domains other than ICT	AI has a wide range of applications outside of computer science. The survey asked participants whether and to what extent AI applications in other domains were considered. Art, music, social studies, science and health were given as examples
Understanding, using and developing AI	Understanding and using AI techniques	This area included (1) the extent to which theoretical understandings of AI processes were developed (e.g. defining or demonstrating patterns, or labeling parts of a machine learning model); and (2) the extent to which students used existing AI algorithms to produce outputs (e.g. training a classifier). Machine learning in general, supervised and unsupervised learning, reinforcement learning, deep learning, and neural networks were given as examples of AI techniques
	Understanding and using AI technologies	AI technologies are often human-facing applications which may be offered "as a service". NLP and computer vision were given as examples. Respondents were asked about the extent to which learners used existing AI technologies to complete tasks or projects, and/or studied the processes of creating these technologies
	Developing AI technologies	Developing AI technologies deals with the creation of new AI applications that may address a social challenge or provide a new type of service. It is a specialized field requiring knowledge of a range of complex techniques and skills in coding, mathematics (especially statistics), and data science

In February 2022, UNESCO published "K-12 AI Curricula: A Mapping of Government-Endorsed AI Curricula", the first report on the global status of K-12 AI education. In this report, UNESCO outlined nine areas of knowledge for K-12 AI education: algorithms and programming, data literacy, contextual problem solving, the ethics of AI, the social or societal implications of AI, applications of AI to domains other than ICT, understanding and using AI techniques, understanding and using AI technologies, and developing AI technologies.